Paracetamol – a curriculum resource

Frank Ellis

ROYAL SOCIETY OF CHEMISTRY

Paracetamol – a curriculum resource

Compiled by Frank Ellis

Edited by Colin Osborne and Maria Pack

Designed by Imogen Bertin

Published by the Royal Society of Chemistry

Printed by the Royal Society of Chemistry

Copyright © Royal Society of Chemistry 2002

Registered charity No. 207890

For further information on other educational activities undertaken by the Royal Society of Chemistry contact:

Education Department
Royal Society of Chemistry
Burlington house
Piccadilly
London W1J 0BA

Email: education@rsc.org
Tel: 020 7440 3344

Information on other Royal Society of Chemistry activities can be found on its websites:
http://www.rsc.org
http://www.chemsoc.org
http://www.chemsoc.org/LearnNet contains resources for teachers and students from around the world.

ISBN 0–85404–375–6

British Library Cataloguing in Publication Data.

A catalogue for this book is available from the British Library.

RS•C

Contents

RS•C

How to use this book

This book consists of seven free-standing activities that can be used singly or as a coherent package in a wide range of teaching and learning situations in both academic and vocational courses. It is aimed at post-16 chemistry students and their teachers.

The book is organised into two main sections – a series of student activities and a guide for teachers and technicians with equipment lists and answers. Within the section for students there are both written and practical tasks. The practical tasks are both preparative/analytical and problem solving. The student sheets can also be freely downloaded from **http://www.chemsoc.org/networks/learnnet/paracetamol.htm**.

Nomenclature

For ease of use some traditional names are retained for chemicals throughout this booklet. Below are listed the traditional names against the systematic names that are commonly used in post-16 chemistry courses in the UK.

Traditional name/trade name	Systematic name
Acetanilide	N-Phenylethanamide
Paracetamol	N-(4-Hydroxyphenyl)ethanamide
Phenacetin	N-(4-Ethoxyphenyl)ethanamide
Phenol	Hydroxybenzene

Health and safety

Teachers must consult their employer's risk assessments before commencing any practical activity. It is good practice to encourage students to do so also. However, this does not absolve teachers from their responsibility to check students' plans and supervise the activity. Eye protection and other appropriate protective equipment should be worn for all the experiments in this booklet. The following texts are available to give guidance on health and safety and assessing risk:

Safeguards in the School Laboratory, 10th edition, ASE, 1996

Topics in Safety, 3rd Edition, ASE, 2001

Hazcards, CLEAPSS, 1995 (or 1998, 2000 updates)

Laboratory Handbook, CLEAPSS, 2001

Safety in Science Education, DfEE, HMSO, 1996

Hazardous Chemicals. A manual for science education, SSERC, 1997 (paper).

Hazardous Chemicals. An interactive manual for science education, SSERC, 2002 (CD-ROM).

Further information can be obtained from:

- CLEAPSS School Science Service at Brunel University, Uxbridge, UB8 3PH; tel 01895 251496; fax 01895 814372; email science@cleapss.org.uk or visit the website **http://www.cleapsss.org.uk** (accessed August 2002).
- Scottish School Equipment Resource Centre (SSERC). Contact SSERC at St Mary's Building, 23 Holyrood Road, Edinburgh, EH8 8AE; tel 0131 558 8180, fax 0131 558 8191, email sts@sserc.org.uk or visit the website **http://www.sserc.org.uk** (accessed August 2002).

Background information

1. Paracetamol is a common compound

The structure of paracetamol

Paracetamol is a very widely used medicine. It is a mild painkiller and reduces the temperature of patients with fever. These actions are known respectively as analgesic and antipyretic.

There are currently more than 90 common products containing paracetamol which are available over the counter from British pharmacies. Many of them are sold as treatments for the relief of cold and influenza and they can be bought in a number of different formulations. Paracetamol is a relatively safe drug but toxic side effects have been observed with high doses greater than 10–15 g.

This toxicity is due to the chemical structure of the compound and the way our bodies break it down. It is metabolised to a reactive intermediate at high doses.

Pure paracetamol is a white crystalline solid which melts at 169–171 °C. Its solubility in cold water is 1.43 g/100 cm^3 but it is much more soluble in hot water (5 g/100 cm^3) and in ethanol (14 g/100 cm^3).

2. The history of paracetamol

At the University of Strassburg in the 1880s Professor Kussmaul, of the Department of Internal Medicine, asked two assistants to give naphthalene as a treatment for intestinal worms.

The medicine had little effect on worms, but one patient had a great reduction in fever temperature. It was found that this patient had, in fact, been given acetanilide instead of naphthalene due to a mistake at the pharmacy!

Acetanilide

Naphthalene

The young assistants quickly published the discovery of this new antipyretic (fever-reducing drug). It was soon in production and remained in use for several years because it was so cheap to produce. However, it had a serious side effect involving the deactivation of some of the haemoglobin in red blood cells.

RS•C

The publication of news about acetanilide immediately spurred a chemist at Bayer's dyeworks to make some derivatives:

N-(4-Methoxyphenyl)ethanamide

N-(4-Ethoxyphenyl)ethanamide
(Phenacetin)

These were both found to be antipyretic and N-(4-ethoxyphenyl)ethanamide was less toxic than acetanilide itself. It was promptly marketed as 'Phenacetin' and has remained in use ever since. However, restrictions have been placed on its use due to kidney damage in long-term users.

Many medicines were synthesised to try to improve on phenacetin and as early as 1893 Joseph von Mering made paracetamol.

N-(4-Hydroxyphenyl)ethanamide
(Paracetamol)

He found it to be an effective antipyretic and analgesic, but wrongly thought that it caused the same haemoglobin problem as acetanilide.

It was not until the 1940s that paracetamol was reinvestigated after it was found present in patients dosed with phenacetin. In 1953 paracetamol was marketed by Sterling-Winthrop Co., and promoted as preferable to aspirin since it was safe to take for children and people with ulcers. However, it causes liver damage from chronic use.

Paracetamol is rapidly formed in the guts of people who take phenacetin. It is the major metabolite (decomposition product) and it is likely that the antipyretic and analgesic effects of phenacetin were in fact due to paracetamol.

There are some suggestions that the toxic effects of phenacetin were due to a minor metabolite – the N-oxide.

Phenacetin is metabolised to two compounds. One involves the removal of the ethyl (CH_3CH_2-) substituent from oxygen. The second involves the replacement of the hydrogen atom on nitrogen by a hydroxyl (-OH) group. This type of compound is called a hydroxamic acid. Hydroxamic acids bind strongly to metal ions. This action may contribute to the toxicity.

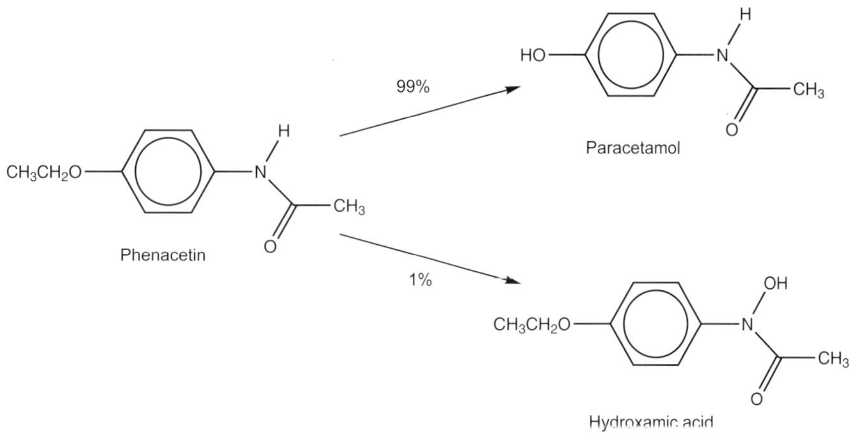
99%
1%
Phenacetin
Paracetamol
Hydroxamic acid

Methods of establishing the safety and efficacy of medicines

At the turn of the 20th century the discovery and testing of medicines was a largely haphazard process with the pioneers of high quality science such as Pasteur and Erhlich being exceptions. As in the history of paracetamol, new compounds were given to patients almost immediately after synthesis or discovery.

In contrast, the development of modern medicines is a long and expensive process typically taking 10-12 years. Chemistry is used at all stages to develop the synthesis and determine purity. Potential patients need to know at least four things about a new medicine before they take it. Does it work? Is it safe? How much do I take? How often do I take it?

The answer to the first question is initially investigated in isolated enzyme or cellular systems before being trialed in animal models and ultimately in volunteer patients. Clinical trials are used in the later stages to see if a new medicine works in one set of patients compared to the effects of a placebo on another group.

The only way to determine safety and dosage regimes are to use animal models before human volunteers take the medicine.

Task
Consider the ethical considerations of testing new medicines on animals.

Experimental and investigative section

The extraction and purification of paracetamol from tablets

This can be used to find out which brand of paracetamol contains the greatest quantity of active ingredient.

Safety

Wear eye protection.

Propanone – volatile, highly flammable, keep away from flames, do not inhale vapour.

Paracetamol tablets – do not ingest.

A full risk assessment should be carried out prior to starting the experiment.

- Warm two paracetamol tablets with propanone (20 cm^3) in a small conical flask by placing the flask in warm water.
- Once the tablets have broken up, the undissolved material (binding agents and filler) should be removed using a filter paper and funnel. Allow the propanone to evaporate, either overnight or on a warm water bath in a fume cupboard.
- The white solid is crude paracetamol. Keep a small amount of the solid to determine its melting point later. The material can be purified by recrystallisation from water.

This process relies on the fact that paracetamol is not very soluble in cold water (1.4 g/100 cm^3) but very soluble in hot water (5 g/100 cm^3). When the crude solid is heated in water it will dissolve and any insoluble impurities can be filtered off. The impurities which are soluble will also of course dissolve. When the hot solution is cooled down, it reaches the temperature at which paracetamol reaches its limit of solubility and therefore starts to crystallise out. However, the soluble impurities are only present to the level of a few percent and so never reach their limit of solubility and thus stay in solution.

- Heat the solid in about 10 cm^3 of water to dissolve it, and filter off any insoluble material through a very small piece of cotton wool in a warm glass funnel. (Pour hot water through the funnel and cotton wool first.)
- Cool the filtrate, and filter off the crystals that form.
- Dry the pure paracetamol by either pressing with filter papers or gently warming in an oven. Take the melting point, and compare it with that of your crude sample and the quoted melting point of pure paracetamol.

Question

Why might the melting points be different?

Paracetamol formulations: presentation activity

Paracetamol can be taken in a number of ways and can be bought in many different formulations. Common ones are tablets (500 mg), fizzy dispersible tablets (500 mg), paediatric oral solutions (120 mg/5 cm^3), oral suspensions (250 mg/cm^3), and suppositories (125 mg). It is also sold in capsules as a mixture with other active ingredients such as codeine and caffeine. Work in small groups and discuss why there might be advantages in having a number of formulations. Comment on the doses available and suggest a target group for each one. Present your findings to the group, perhaps as a poster or as an audio-visual presentation *eg* Powerpoint presentation.

Hints on presentations
In the presentation you could include the following:

- the conditions that paracetamol helps to relieve or cure, including technical terms such as analgesic, antipyretic and anti-inflammatory;
- the side effects of paracetamol, and the alternative treatments for people who are affected by them;
- the historical development of paracetamol, including the achievements of those responsible for the main developments;
- the chemistry involved in developing the medicine in a usable form; and
- the nature and importance of clinical trials.

You may find information in reference books, in libraries, in pharmacies and by contacting the ABPI (Association of the British Pharmaceutical Industry, 12 Whitehall, London, SW1A 2DY, **http://www.abpi.org.uk** Tel: 020 7930 3477) or pharmaceutical companies.

A recommended website is **http://www.pharmweb.net/pwmirror/pwy/paracetamol/pharmwebpic.html** (accessed August 2002).

Making a poster
In making a poster the following hints may be useful:

- your poster should be clearly set out, the structure should be clear at a glance;
- people do not like reading a lot of text. Diagrams and flow charts are much easier to take in; text should be readable from at least 2 m;
- explanations should be separate from the main story, perhaps in distinctive boxes; and
- the level must be appropriate for the expected audience: you will need to think about what the audience is likely to know already.

Making an audio-visual presentation
In making an audio-visual presentation the following hints may be useful:

- before you start, make sure you have everything ready and you know how to switch on the OHP or operate the data projector and that it is focussed correctly;
- start the presentation with something designed to capture attention and to help your audience to know what to expect;
- do not read directly from notes: use notes if you need to, but always talk directly to your audience;
- people get bored if they have nothing to do but listen to you talking: make sure that there is always something to look at as well;
- make sure your visual aids are prepared well beforehand: they are a very effective way of getting information across to your audience;
- if you are drawing formulae on a white board or black board make sure that you know them by heart (draw them out beforehand): you should not have to keep looking at your notes to make sure that you have got something right;
- remember that you are always more familiar with your subject matter than your audience: give them time to take in what you are saying before going on to the next stage; and
- mannerisms are irritating, so try to stand still, look at your audience and do not wave your hands about, or keep scratching your nose or trip over the OHP lead!

The preparation of paracetamol

Paracetamol can be made in three steps from phenol

dilute H_2SO_4

NaNO$_3$

25% yield   36% yield

Step 1

NaBH$_4$

Pd / 1 mol dm^{-3} NaOH

74% yield

Step 2

Ethanoic anhydride

Water, Room temp.

Step 3

Step 1 – the nitration of phenol

Preparing 4-nitrophenol and 2-nitrophenol

Safety

Wear eye protection. Protective gloves should be worn when handling phenol.

Sodium nitrate(V) – oxidising agent, keep away from flammable material.

Concentrated sulfuric acid – very corrosive dense liquid, dehydrating agent, add very slowly with constant stirring to water whilst cooling.

Phenol – toxic by ingestion and skin absorption. It can cause severe burns. Take care when removing phenol from the bottle because the solid crystals can be hard to break up. Wear protective gloves and work in a fume cupboard.

2- and 4-nitrophenols – harmful and irritant, will stain skin, wear protective gloves when handling.

1. Place a 3-necked round bottom flask in an ice water bath and place a thermometer into one of the necks. Place 15 g of sodium nitrate(V) into the flask, add 40 cm^3 of water and stir.

2. Cautiously add concentrated sulfuric acid (13.6 cm^3; 25 g) to the stirred solution.

3. Slowly add solid phenol (9.4 g , 0.1 mol) at such a rate that the temperature of the solution does not rise above 20 °C (about a half a spatula at a time over about 20 minutes) and then stir, preferably with a magnetic stirrer, for 2 hours.

4. Remove the thermometer. Decant off the supernatant liquid and add water (25–30 cm^3) to the residue. Put a dropping funnel into one of the necks and a stillhead and condenser into another. Insert a stopper in the third.

5. Heat the mixture, and distil off one of the components with the steam. At the same time add water to the mixture through the dropping funnel at a similar rate.

6. Stop the distillation once the product has stopped coming over, then filter the distillate to give crystals of the 2-nitrophenol isomer.

7. Cool the residual solution in the distillation flask, then filter off the other solid isomer (4-nitrophenol) and recrystallise it from 0.5 mol dm^{-3} hydrochloric acid.

Questions

1. Why does the reaction give a mixture of isomers? What other product might you get?

2. Why does the nitration need only mild conditions, whilst benzene requires concentrated H$_2$SO$_4$/HNO$_3$?

3. What happens when you add concentrated sulfuric acid to sodium nitrate(V)?

4. The products can be separated by a simple steam distillation procedure. The 2-nitrophenol product is volatile in steam but the 4-nitrophenol is not. Why?

5. The two products can be separated by chromatography. The 2-nitrophenol has less affinity for silica (*ie* runs faster) than the 4-nitrophenol. Why?

6. What are the melting points of the pure products? Why is one lower than the other?

7. Dissolve each product in water - which solution is more acidic and why?

8. Look up the melting points of the following pairs of compounds (be careful how you interpret the names).
3-methyl-2-nitrophenol
3-methyl-4-nitrophenol
5-methyl-2-nitrophenol
3-methyl-4-nitrophenol
5-fluoro-2-nitrophenol
3-fluoro-4-nitrophenol

What do you notice about the melting points?

Step 2 – the reduction of a nitro group to an amine

HO—⟨benzene ring⟩—NO₂ →(NaBH₄, Pd / 1 mol dm⁻³ NaOH)→ HO—⟨benzene ring⟩—NH₂

74% yield

Preparing 4-aminophenol

Safety

Wear eye protection.

Sodium tetrahydridoborate(III) – harmful if swallowed, reacts with water to produce hydrogen; the reaction is more vigorous with acids – keep away from flames.

4-nitrophenol – irritant, will stain skin, wear gloves when handling.

Sodium hydroxide solution (1 mol dm⁻³) – corrosive – avoid contact with eyes.

Palladium on charcoal (5% or 10%) – Irritant. Do not breathe dust.

Hydrochloric acid (2 mol dm⁻³) – irritant.

Sodium hydrogen carbonate – reacts with acids to produce carbon dioxide. The reaction can be delayed and violent. Add cautiously with constant agitation and hold the flask over a dish to catch any spill-over.

4-aminophenol – harmful, possible risk of irreverisble effects, very toxic to aquatic organisms.

Note – This reaction needs careful temperature control.

1. Place 10 cm³ (10 mmol) of 1 mol dm⁻³ sodium hydroxide in a conical flask.
2. Add 0.56 g (14.7 mmol) of sodium tetrahydridoborate(III) (sodium borohydride), followed by 50 mg of palladium on charcoal (5% or 10%; Aldrich).
3. Cool in ice to ~13 °C.
4. Add 1.0 g (7.2 mmol) of 4-nitrophenol in very small portions (half a microspatula at a time) over 30 minutes. Make sure the temperature is kept between 13–17 °C during the addition.
5. After the addition is complete the mixture should be stirred for a further 15 min and acidified with 2 mol dm⁻³ hydrochloric acid (about 17 cm³).
6. Filter the mixture to remove catalyst and adjust the filtrate to pH 7–8 by carefully adding solid sodium hydrogencarbonate a little at a time.
7. Filter off the precipitate and wash with a little cold water to give 4-aminophenol (0.58 g; 74%) after drying.

Questions

1. Sodium tetrahydridoborate(III) (sodium borohydride), $NaBH_4$ is relatively stable in aqueous sodium hydroxide (NaOH) but not in acid. Why?
2. What is the role of the catalyst in the reduction of nitro groups? Can other methods be used for this reduction?
3. Why can the product be separated from unreacted starting material at pH 8?
4. Why is sodium hydrogencarbonate used to make the reaction mixture basic at the end of the reaction, rather than sodium carbonate or sodium hydroxide?
5. Why is the product soluble in a solution of a strong acid or in a solution of a strong base, but not in a solution of a weak base?

Step 3 – the formation of an amide

Preparing *N*-(4-hydroxyphenyl)ethanamide – Paracetamol

Safety

Wear eye protection.

4-aminophenol – irritant.

Ethanoic anhydride – corrosive, causes burns, flammable, the vapour will irritate the eyes and the respiratory system. Use in a fume cupboard.

Paracetamol – do not ingest.

1. Place 1.0 g of 4-aminophenol and 9 cm^3 of distilled water in a 50 cm^3 conical flask and stir briskly at room temperature, in order to suspend the solid in the water.
2. In a fume cupboard, add 1.1 cm^3 (1.17 g) of ethanoic anhydride to the stirred suspension and gently shake to mix. The solid will dissolve after about 30 seconds. Continue shaking and a precipitate will form after 2 minutes.
3. After 10 minutes the solid should be filtered off under suction, washed with a little cold water and dried (0.83g; 60%).
4. The product may be purified by crystallisation from distilled water. Dissolve the crude product in the minimum of distilled water at about 80 °C (you will probably require about 15 cm^3).
5. Allow the clear solution to cool slowly to room temperature and collect the recrystallised product by suction filtration, washing with 5 cm^3 of ice-cold distilled water.
6. Dry the recrystallised product either between filter papers or by gently warming in an oven, and determine the yield.
7. Determine the melting point of the dry, recrystallised product.

The melting point should be 169–171 °C.

Questions

1. Why is the product insoluble in aqueous ethanoic acid, but the starting material is soluble?
2. If the reaction is done in dilute hydrochloric acid rather than water, the product is

Explain why this is.

The quantitative analysis of various formulations of paracetamol

The British Pharmacopoeia method for the analysis of paracetamol involves heating it under reflux with $1\,mol\,dm^{-3}$ sulfuric acid. This is a straightforward, acid catalysed, hydrolysis of an amide to an amine and a carboxylic acid. The 4-aminophenol which is formed is then titrated with an oxidising agent, ammonium cerium(IV) sulfate using ferroin as the indicator.

The first reaction is as follows:

4-aminophenol   Ethanoic acid

The titration step is much more interesting. 4-Aminophenol can easily be oxidised as follows:

4-aminophenol   Iminoquinone

The role of the ammonium cerium(IV) sulfate is to oxidise the 4-aminophenol to the iminoquinone. Only after all the 4-aminophenol has been oxidised will the cerium (IV) reagent oxidise the ferroin indicator from Fe^{2+} to Fe^{3+} (ferriin).

Ferroin (red)   Ferriin (blue)

During the titration the solution should be red, and the yellow end point is the transition from red to pale blue.

It is easy to work out that, since 1 mole of Ce^{4+} is equivalent to 0.5 mole of paracetamol, the conversion factor given in the method is correct.

RS•C

Quantitative analysis of paracetamol – page 1 of 2

Procedure as outlined in the British Pharmacopoeia 1988

Safety

Wear eye protection.

Paracetamol formulations – do not ingest.

Sulfuric acid (1 mol dm^{-3}) – corrosive, especially when hot.

Hydrochloric acid (2 mol dm^{-3}) – irritant.

Ferroin solution – hazards unknown. May cause skin irritation.

Ammonium cerium(IV) sulfate – respiratory tract irritant, strong oxidising agent, keep away from flammable material.

1. Dissolve 0.3 g of a mixture containing paracetamol in a mixture of water (10 cm^3) and 1 mol dm^{-3} sulfuric acid (30 cm^3).
2. Boil under reflux for 1 hour, cool and dilute with water (100 cm^3).
3. To 20 cm^3 of the resulting solution add cold water (40 cm^3, 2 mol dm^{-3} hydrochloric acid (15 cm^3) and ferroin solution (0.1 cm^3, 0.1 wt% or 0.025 mol dm^{-3}).
4. Titrate with 0.1 mol dm^{-3} ammonium cerium(IV) sulfate (VS – volumetric standard) until a yellow colour is produced.
5. Repeat the operation without the test material being present. The difference between the titration figures represents the amount of ammonium cerium(IV) sulfate required. Each cm^3 of 0.1 mol dm^{-3} ammonium cerium(IV) sulfate is equivalent to 0.007560 g of paracetamol.

This method can be used to analyse the quantity of paracetamol present in many medicines that contain the drug.

Question

How could you prove that the first step in the quantitative analysis of paracetamol involves hydrolysis to 4-aminophenol?

Using thin-layer chromatography to investigate the reactions

You have probably used a simple chromatography experiment as part of your earlier studies to separate the dyes in a coloured ink. The same technique can be used to separate substances which are not dyes but in such experiments the chromatogram must be developed to show up the various different substances that have been separated.

Chromatography techniques are used a great deal in industry because they can be controlled very precisely and use very small amounts of substance. In this activity you investigate the purity and identity of your laboratory prepared samples of nitrophenol or paracetamol using thin-layer chromatography (tlc). In this activity all the substances are white or pale yellow so you will need to develop the plate before you can see what has happened.

Thin-layer chromatography is a powerful tool for determining if two compounds are identical. A spot of the compound being investigated is placed on a chromatography plate, and a spot of a pure manufactured sample of the same substance is placed next to it. The plate is then allowed to stand in a suitable solvent, which travels up the plate. If the compound to be identified leaves exactly the same pattern on the chromatography plate as the known pure compound it is reasonable to conclude that they are the same. However, if extra spots are observed as well as the characteristic pattern of the known compound, then impurities are likely to be present in the sample.

In the experiment both crude samples of 2-nitrophenol and 4-nitrophenol are compared with known samples.

Safety

Wear eye protection.

Ethyl ethanoate – volatile, highly flammable, keep away from flames, irritant, do not inhale vapour.

Cyclohexane – volatile, highly flammable, keep away from flames, harmful, do not inhale vapour.

Iodine – harmful by inhalation or skin contact.

Short wave UV – may cause skin cancer and eye damage. Do not observe directly. The viewer should be screened from direct radiation.

1. Make sure that you do not touch the surface of the tlc plate with your fingers during this activity. Handle the plate only by the edges and use tweezers if possible.
2. Take a tlc plate and using a pencil (not a ball-point or felt tip pen) lightly draw a line across the plate about 1 cm from the bottom. Mark four equally spaced points on this line.
3. Place small amounts (about 1/3 of a spatula measure) of your crude 2-nitrophenol, your crude 4-nitrophenol and the commercial samples of these in four separate test-tubes. Label the test-tubes so that you know which is which.

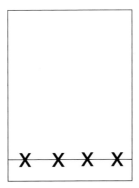

4. Add 1 cm³ of ethyl ethanoate to the test-tubes to dissolve the samples. If possible do this in a fume cupboard.
5. Use clean capillary tubes to spot each of your four samples onto the tlc plate. Allow the spots to dry and then repeat. The spots should be about 1–2 mm in diameter.
6. After all the spots are dry, place the tlc plate in the chromatography tank making sure that the original pencil line is above the level of the developing solvent – ethyl ethanoate:cyclohexane 1:4. Put a lid on the tank and allow to stand in a fume cupboard until the solvent front has risen to within a few millimetres of the top of the plate.

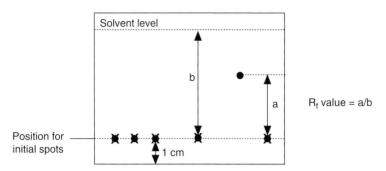

Solvent level

b

a

R_f value = a/b

Position for initial spots

1 cm

7. Remove the plate from the tank and quickly mark the position of the solvent front. Allow the plate to dry.

8. Observe the plate under a short wavelength UV lamp and lightly mark with a pencil any spots observed.

9. Carefully place the plate in a jar or beaker containing a few iodine crystals. Put a cover on the jar and wait for the spots to appear. Do this in a fume cupboard if possible.

Results
■ Draw a diagram to show which spots appeared under UV light and which appear with iodine.

■ Determine the R_f value of the samples using the expression

R_f = distance moved by sample / distance moved by solvent

Questions
1. Write a short paragraph explaining why some substances move further up the tlc plate than others and how the results are made visible.

2. What conclusions can you draw about the nature of the four samples tested?

Other experiments
The conversion of 4-aminophenol into paracetamol can be followed using ethyl ethanoate:cyclohexane 2:1 as the eluant.

This system can also be used to monitor the hydrolysis step in the quantitative analysis.

To test aqueous mixtures, take a small sample (less than 1 cm^3) and place in a test-tube. Adjust the solution to pH 8 by adding small amounts of sodium hydrogen carbonate. Add ethyl ethanoate (1 cm^3) and shake. Allow the layers to separate and use a capillary tube to sample the upper ethyl ethanoate layer for tlc.

Control of the pH of the aqueous sample is important to ensure the organic compound is not ionised otherwise it will not extract effectively or run on a tlc plate. Use known samples of 4-aminophenol and paracetamol as reference samples on the plate.

This could form a separate investigation to semi-quantitatively look at aqueous pH versus how much can be extracted with ethyl ethanoate. Look at the spot size after running a tlc. Investigate paracetamol, 4-aminophenol and 4-nitrophenol.

Appendix

Column chromatography – the separation of 2- and 4-nitrophenols

Safety

Wear eye protection.

Use a column with a sintered glass disc.

Ethyl ethanoate – volatile, highly flammable, keep away from flames, irritant, do not inhale vapour.

Cyclohexane – volatile, highly flammable, keep away from flames, harmful, do not inhale vapour.

Silica – its dust is a respiratory tract irritant.

Petroleum ether 60–80 °C – harmful, highly flammable

(It is possible to use solid phase extraction cartridges (10 g silica) and reduce the scale)

A full risk assessment should be carried out prior to starting the experiment.

Add sand to the disc to a depth of 5–10 mm, and add the solvent mixture to fill the column to about one-third. The solvent is ethyl ethanoate:cyclohexane 1:4 (petroleum ether 60–80 °C can be used instead of cyclohexane)

Weigh out silica (50 g) (in a fume cupboard and wearing gloves) and add enough solvent to give a mobile (runny) slurry. Pour this gently into the column with the tap open. Run the solvent down to the top of the silica while tapping the column a few times to pack the particles closely in it. Gently add another 5–10 mm layer of sand to the top of the material in the column.

Dissolve a mixture of the nitrophenols (0.25 g of each) in ethyl ethanoate (0.25 cm^3) and dilute with cyclohexane (2 cm^3). Carefully apply the solution to the sand on top of the column. Run the mixture onto the silica by opening the tap slightly and gently adding solvent. Once the phenols are on the silica and off the sand, fill the column with solvent, adding it very slowly at first so as not to disturb the top layer.

Start running the column by opening the tap, using a modest drip rate, and collect fractions in test-tubes. Keep topping up the solvent head at this point.

Check each fraction by tlc using the same solvent mixture, and combine the fractions which contain the same component of the mixture. Evaporate off the solvent, by leaving the solution in a small open vessel in a fume cupboard.

RS•C

Teachers' notes

Presentation activity

Paracetamol can be taken in a number of ways. Tablets (500 mg) – the simplest form of taking a medicine for most adults – 1 or 2 tablets is well below the level at which there are toxic effects.

Fizzy dispersible tablets (500 mg) – for adults who have trouble swallowing tablets. The paracetamol is mixed with citric acid and sodium hydrogencarbonate which effervesce when put into water. The effervescence rapidly disperses the drug into a fine suspension.

Paediatric oral solutions (120 mg/5 cm^3) – for young children who cannot take tablets – note that the recommended dose is a quarter of the adult dose. Also note that this concentration is higher than the normal aqueous solubility – this can be achieved by adding a little ethanol.

Oral suspensions (250 mg/5 cm^3) – for older children and adults who dislike tablets. It has to be a suspension at this concentration as it will not dissolve sufficiently.

Suppositories (125 mg) – drugs are rapidly absorbed from the colon and metabolism is significantly reduced because the drug does not have to go via the stomach, therefore the dose can be lower.

Capsules (various dosages) – the gelatin coating is 'slippery' and the cylindrical shape makes the capsules easier to swallow than tablets. Also, it is a simple way of delivering a mixture of solids.

RS•C

The toxicity of paracetamol – information sheet

Why is paracetamol toxic at high doses?

Paracetamol is metabolised to some extent by oxidation to an iminoquinone in a way similar to the chemistry described in the quantitative analysis section. Although the iminoquinone is a very reactive compound, it is rapidly deactivated and converted into a highly water soluble derivative by the addition of endogenous glutathione – a small peptide with a thiol (–SH) group. This 'conjugation' by glutathione is a very common elimination pathway of the body. The reaction makes 'foreign' compounds so water soluble that they are excreted in the urine.

Our bodies only have a limited amount of glutathione available at any one time and so an overdose of paracetamol rapidly depletes this protective chemical. This depletion allows the iminoquinone and other toxins to build up and seriously damage our cells.

With a knowledge of this chemistry it is easy to understand why the antidote for paracetamol overdose is acetylcysteine, a simple non-toxic thiol which can deactivate the iminoquinone. Acetylcysteine is also a precursor for glutathione synthesis.

RS•C

Apparatus lists and answers

The extraction and purification of paracetamol from tablets

Chemicals (per group)
- Two paracetamol tablets
- Propanone

Apparatus and equipment (per group)
- Two conical flasks
- Measuring cylinder
- Warm water bath
- Filter funnel and filter papers
- Beaker
- Pipette
- Small piece of cotton wool
- Glass funnel
- Hot and cold water
- Bunsen burner, tripod and gauze
- Access to melting point apparatus
- Access to a balance
- Eye protection

This activity could be set as an open ended investigation.

Answer
The initial crude sample contains impurities or ingredients that were not removed in the extraction process. A purer compound has a higher melting point. Impurities present in compounds lower the melting point by affecting the crystal structure, so the crystals do not pack together as well.

The preparation of paracetamol
Step 1 – the nitration of phenol

Preparing 2-nitrophenol and 4-nitrophenol

Chemicals (per group)
- Concentrated sulfuric acid
- Sodium nitrate(V)
- Solid phenol
- Hydrochloric acid (0.5 mol dm^{-3})

Apparatus and equipment (per group)
- 3-necked round bottom flask
- Magnetic stirrer
- Ice bath
- Thermometer

RS•C

- Quickfit® distillation apparatus (still head, condenser, stoppers)
- Dropping funnel
- Spatulas
- Access to a balance
- Access to a fume cupboard

Answers

1. It is possible to get 2- and 4- substitution of the ring due to effective charge stabilisation at both these sites. The –OH group activates the ring preferentially at the 2- and 4- positions and therefore directs substituents to these positions (rather than to the 3- and 5- positions which are not activated). Another product could be the di-substituted form:

2. Phenol activates the ring to electrophilic attack because the lone pair on the oxygen pushes electron density onto the ring. This means nitration is easier and does not require such harsh conditions.

3. $H_2SO_4 + 2NaNO_3 \rightarrow Na_2SO_4 + 2HNO_3$
   Then in excess sulfuric acid: $HNO_3 + H_2SO_4 \rightarrow NO_2^+ + HSO_4^- + H_2O$
   This is sometimes seen as: $HNO_3 + 2H_2SO_4 \rightarrow NO_2^+ + 2HSO_4^- + H_3^+O$

4. 4-Nitrophenol is better at hydrogen bonding with other molecules of 4-nitrophenol and/or water so it is less volatile. 2-nitrophenol has intramolecular hydrogen bonding and consequently, fewer intermolecular hydrogen bonds with other molecules, so is more volatile.

5. 2-Nitrophenol is less polar due to intramolecular hydrogen bonding.

6. Melting point of pure 4-nitrophenol = 113.8 °C
   Melting point of pure 2-nitrophenol = 45.1 °C

   4-Nitrophenol has a higher melting point due to intermolecular hydrogen bonding, holding the molecules together in the crystal.

7. 4-Nitrophenol is more acidic. The hydrogen on the OH of the 2-nitrophenol is stabilised by intramolecular hydrogen bonding and so it is not released as readily.

8. This question gives three examples of pairs of nitro phenols where the 2-isomer has a lower melting point than the 4-isomer. The melting points can be found in a general chemical supplies catalogue such as Aldrich – Handbook of Fine Chemicals and Laboratory Equipment.
   3-methyl-2-nitrophenol m.p. 35–39 °C
   3-methyl-4-nitrophenol m.p. 127–129 °C
   5-methyl-2-nitrophenol m.p. 53–56 °C
   3-methyl-4-nitrophenol m.p. 127–129 °C
   5-fluoro-2-nitrophenol m.p. 35–37 °C
   3-fluoro-4-nitrophenol m.p. 93–95 °C

It is an interesting exercise in seeing how names and locants change in that 3- and 5-substituents are in fact the same position with respect to the phenol -OH.

RS•C

Other examples that can be found in the Dictionary of Organic Compounds (published by Chapman & Hall):

3-chloro-2-nitrophenol m.p. 37–38 °C

3-chloro-4-nitrophenol m.p. 121–122 °C

5-chloro-2-nitrophenol m.p. 42–43 °C

5-methoxy-2-nitrophenol m.p. 95–96 °C

3-methoxoy-4-nitrophenol m.p. 144–145 °C.

Step 2 – the reduction of a nitro group to an amine

Preparing 4-aminophenol

Chemicals (per group)
- 4-Nitrophenol
- Sodium tetrahydridoborate(III) (sodium borohydride)
- Sodium hydroxide (1 mol dm^{-3})
- Palladium on charcoal (5% or 10%, Aldrich)
- Hydrochloric acid (2 mol dm^{-3})
- Sodium hydrogencarbonate

Apparatus and equipment (per group)
- 2 Wide mouthed conical flasks or beakers
- Magnetic stirrer
- Thermometer
- Spatula
- Measuring cylinder
- Filter paper and funnel
- Water pump
- Buchner flask and funnel
- Microspatula
- Eye protection

Answers
1. In acid solution you would get hydrogen gas released, $H^+ + H^- \rightarrow H_2$
2. The catalyst provides a surface for the reaction to take place on by forming weak association bonds with the catalyst surface. This in turn weakens the strong nitro bonds. Other methods that could be used are:
   – Sn/conc HCl followed by $NaHCO_3$
   – H_2/Pd
   – $LiAlH_4$
3. NO_2 group in the starting material makes the phenol OH more acidic than in the product (NH_2 is less electron withdrawing) and therefore dissolves in the basic solution while the product precipitates out.
4. Sodium carbonate and sodium hydroxide are strong enough bases to remove the H^+ from the products as well. Sodium hydrogencarbonate allows differentiation between product and starting material.
5. In strong acid or base a salt is produced that is water soluble. (A strong acid will protonate the $-NH_2$ to give $-NH_3^+$, *ie* soluble; a strong base will deprotonate $-OH$ to give $-O^-$ also soluble.)

Step 3 – the formation of an amide

Preparing *N*-(4-hydroxyphenyl)ethanamide – Paracetamol

Chemicals (per group)
- 4-Aminophenol
- Ethanoic anhydride

Apparatus and equipment (per group)
- Two 50 cm^3 conical flasks
- One 100 cm^3 beaker
- Tripod and gauze
- Bunsen Burner
- Thermometer (-10 °C to 110 °C)
- Buchner flask, funnel and filter papers
- Melting point apparatus
- Glass rod or magnetic stirrer
- One 10 cm^3 and one 50 cm^3 measuring cylinder
- Clamp and stand
- Access to oven and fume cupboard
- Ice and water supply
- Water pump
- Eye protection

Answers

1. Amides are not basic because the lone pair on nitrogen is delocalised over the carbonyl group. They are therefore not soluble in acid. Amines are basic and do dissolve in acid.

2. At low pH the NH_2 is protonated and allows the OH to react with ethanoic anhydride. The ester that is produced is still soluble in the acid due to $-NH_2$ protonation, but due to partial dissociation, this group is eventually ethanoylated and thrown out of the solution. At higher pH, the NH_2 is not protonated and rapidly gets ethanoylated and thrown out of solution before the –OH has a chance to react.

The quantitative analysis of various formulations of the reactions

Chemicals
- Sulfuric acid (1 mol dm^{-3})
- Hydrochloric acid (2 mol dm^{-3})
- Ammonium cerium(IV) sulfate (0.1 mol dm^{-3})
- Paracetamol
- Water
- Ferroin solution (0.1% wt or 0.025 mol dm^{-3})

Apparatus and equipment (per group)
- Round bottom flask
- Quick fit reflux apparatus
- Heating mantle, or Bunsen burner, tripod and gauze
- Clamps and stand
- Spatula

RS•C

- One 10 cm^3 measuring cylinder
- One 50 cm^3 measuring cylinder
- Burette
- Two conical flasks
- Beaker
- Eye protection

Procedure as outlined in the British Pharmacopoeia 1988

Answer
Isolate the material and characterize it, determine the melting point and compare it with the previous one, also tlc, or follow the reaction by running tlc plates using paracetamol and 4-aminophenol as reference spots.

Using thin-layer chromatography to investigate the reactions

Chemicals (per group)
- 2-Nitrophenol
- 4-Nitrophenol
- Ethyl ethanoate
- Cyclohexane
- Iodine
- Solid sodium hydrogencarbonate

Apparatus and equipment (per group)
- Four test-tubes
- Marker pen and pencil
- Access to fume cupboard
- Capillary melting point tubes
- Tlc plate (or chromatography no. 1 paper)
- Chromatography tank (or 1 dm^3 beaker)
- Measuring cylinders
- Spatula
- Access to UV lamp screened from direct view
- Eye protection

Answers
1. Less polar substances move further up the plate because they have less affinity for polar silica – results made visible using ultraviolet light, iodine staining or potassium permanganate staining.
2. Dependent on results.

RS•C